An Alternative Solution to Fermat's Last Theorem

Paul Mansbridge

I0494171

2000 *Mathematics Subject Classification* 00A05 (primary) 03C05 (secondary).

ABSTRACT

An alternative proof is provided for the statement "$x^n + y^n = z^n$ has no integer solutions for any integer $n > 2$". This proof uses the principle of ratios between integer pairs rather than the use of elliptic curves included in an alternative proof.

1. Introduction

In 1637 the French mathematician Pierre de Fermat stated that he had found a proof for the statement "$x^n + y^n = z^n$ has no integer solutions for any integer $n > 2$" but he also stated that he had insufficient space to write it down in the margin of his copy of "Arithmetica" by Diophantus [1]. No one succeeded in proving this theorem until 1995 when Andrew Wiles published a proof that includes the use of elliptic curves [2] [3]. The following proof uses an alternative method involving ratios between integer pairs.

Let us define a set $\{2^n, 2^{n+1}, 2^{n+2}, ...2^{n+\infty}\}$ as $2^{n...\infty}$ where each value represents 2 raised to a successive power $\geq n$. Denote the set $\{5^n, 5^{n+1}, 5^{n+2}, ...5^{n+\infty}\}$ as $5^{n...\infty}$ where each value represents 5 raised to a power $\geq n$. Denote the set $\{10^n, 10^{n+1}, 10^{n+2}, ...10^{n+\infty}\}$ as $10^{n...\infty}$ where each value represents 10 raised to a power $\geq n$.

There are two parts to this proof. In sections 2 and 3 it is demonstrated that all integer pairs have an nth root that is a whole or fractional multiple of the nth root of another integer pair (e, f) where $e \in 10^{1...\infty}$ and/or $f \in 10^{1...\infty}$. In sections 4 and 5 it is demonstrated that, for any two integers (a, b), when $n > 2$, $(a^n - b^n) \notin 10^{1...\infty}$. Thus, adding any value $\in 10^{1...\infty}$ to an integer raised to a power $n > 2$ results in a total with an irrational nth root, therefore $\sqrt[n]{e^n + f^n}$ is irrational $\rightarrow$ for any integer pair (x, y) with the same ratio as (e, f) $\sqrt[n]{x^n + y^n}$ is irrational.

1

2. Terminology and theorems

A remainder after division is referred to by 'mod'. Let us denote two integers x, y. When $(x \times y) \in 10^{1\cdots\infty}$, x will be referred to as 'compatible', that is $x \in \{ 1, 2^{1\cdots\infty}, 5^{1\cdots\infty}, 10^{1\cdots\infty}, (2^{1\cdots\infty} \times 10^{1\cdots\infty}), (5^{1\cdots\infty} \times 10^{1\cdots\infty}) \}$. Thus y is also compatible. All compatible numbers > 1 end in 0, 2, 4, 5, 6 or 8.

THEOREM 2.1. *Any integer pair can be multiplied by a whole or fractional number ≥ 1 to form a ratio in which one or both sides have a value $\in 10^{1\cdots\infty}$.*

THEOREM 2.2. *An integer pair, when raised to any power $n \geq 1$, has a difference, sum and nth root of the sum that are a whole or fractional multiple of those of another integer pair that is a whole or fractional multiple of the first pair.*

THEOREM 2.3. *Where $a > b$ and $n > 1$, $(a - b) \mid (a^n - b^n)$.*

THEOREM 2.4. *For even powers, $(a^{n/2} - b^{n/2}) \mid (a^n - b^n)$ and $(a^{n/2} + b^{n/2}) \mid (a^n - b^n)$.*

THEOREM 2.5. *The last digit of an integer has a 2 part cycle for each successive odd power.*

2.1. Lemma 1: Irrational roots

An integer with no integer root for any power $n > 1$ has instead a non-recurring decimal root that forms an irrational number. It will be proved in sections 2 and 3, relating to odd and even powers respectively, for two integers (a, b, where $a > b$), that $(a^n - b^n) \notin 10^{1\cdots\infty}$ when $n > 2$. Thus, adding any value $\in 10^{1\cdots\infty}$ to an integer raised to a power $n > 2$ results in a total with an irrational nth root.

Subsection 3.1 shows that any integer pair (x, y) can be multiplied by a whole or fractional number to form (e, f) where $e \in 10^{1 \cdots \infty}$ and/or $f \in 10^{1 \cdots \infty}$. It will also show that, when $n > 2$, $\sqrt[n]{e^n + f^n}$ is irrational $\rightarrow \sqrt[n]{x^n + y^n}$ is irrational since $\sqrt[n]{x^n + y^n} = \frac{x}{e} \times \sqrt[n]{e^n + f^n}$.

3. Proof of theorems

3.1. Proof of Theorem 2.1: Ratios of pairs

Let us assume the integers x, y, r where $x \geq y$. (x, y) can be multiplied by a whole or fractional number ≥ 1 to form a ratio of $r : 10^n$ where $\frac{r}{10^n} = \frac{x}{y}$ and $n \geq 1$. Let us define an integer pair (e, f) where e represents the left side of the ratio and f represents the right side, in this scenario $e = r$ and $f = 10^n$.

For example: $x = 1, y = 1, r = 10$, $(1,1) \rightarrow 10{:}10$, $e = 10, f = 10$.

For example: $(9, 8) \rightarrow 1125{:}1000$.

The ratio may include two infinite integers, including 10^∞, for example: $(5, 3)$ becomes $(5 * [3 \text{ recurring}], 3 * [3 \text{ recurring}]) \rightarrow 1\dot{6} : 10^\infty$, that is ($[1 \text{ followed by } 6 \text{ recurring}] : 10^\infty$).
The ratio can also be expressed as $10^n : r$ where $\frac{10^n}{r} = \frac{x}{y}$. In this scenario $e = 10^n$ and $f = r$.

For example:
$(4, 3) \rightarrow 100{:}75$, $e = 100, f = 75$.

3.2. Proof of Theorem 2.2: Multiples of pairs

Let us denote two integers (a, b) where $a \geq b$, with a whole or fractional common divisor $x > 1$. When a and b are raised to any power $n \geq 1$, a^n and b^n have a difference, sum and nth root of the sum that are a multiple of those of the 'underlying' pair (c, d) that represents the first pair divided by x, that is $c = \frac{a}{x}, d = \frac{b}{x}$.

For example:
Let us denote $a = 55, b = 33, x = 11$, therefore $c = 55/11 = 5, d = 33/11 = 3$.

$$a^n = (x \times c)^n = x^n \times c^n \tag{1}$$
$$b^n = (x \times d)^n = x^n \times d^n \tag{2}$$
$$a^n - b^n = (x^n \times c^n) - (x^n \times d^n) = x^n \times (c^n - d^n) \tag{3}$$
$$a^n + b^n = (x^n \times c^n) + (x^n \times d^n) = x^n \times (c^n + d^n) \Rightarrow \tag{4}$$
$$\sqrt[n]{a^n + b^n} = x \times \sqrt[n]{c^n + d^n} \tag{5}$$

For example:
$a = 12, b = 8, x = 4, c = 3, d = 2.$

$$12^2 - 8^2 = (4^2 \times 3^2) - (4^2 \times 2^2) = 4^2 \times (3^2 - 2^2) = 80 \qquad (6)$$
$$12^2 + 8^2 = (4^2 \times 3^2) + (4^2 \times 2^2) = 4^2 \times (3^2 + 2^2) = 208 \qquad (7)$$
$$\sqrt[3]{12^3 + 8^3} = 4 \times \sqrt[3]{3^3 + 2^3} \qquad (8)$$

3.3. Proof of Theorem 2.3: Where $a > b$ and $n > 1$, $(a - b) \mid (a^n - b^n)$

$(a^n - b^n) = (a \times (a^{n-1} - b^{n-1})) + ((a - b) \times b^{n-1}).$
$(a^2 - b^2) = (a \times (a^1 - b^1)) + ((a - b) \times b^1)$, therefore $(a - b) \mid (a^2 - b^2)$.
In this way, for each power n in turn, it is proved that $(a - b) \mid (a^n - b^n)$,
since $(a^n - b^n)$ is the sum of (1) a multiple of $(a^{n-1} - b^{n-1})$ plus (2) a
multiple of $(a - b)$.

For example:
$7^3 - 4^3 = ((7^2 - 4^2) \times 7) + ((7 - 4) \times 4^2)$
$\quad = ((49 - 16) \times 7) + ((7 - 4) \times 16) = 279 = 93 \times (7 - 4).$
Note: It will be significant that $(a^n - b^n) > (a \times (a^{n-1} - b^{n-1})) \rightarrow$
$(a^n - b^n) > 2$ when $n > 1$.

3.4. Proof of Theorem 2.4: For even powers, $a^{n/2} - b^{n/2} \mid (a^n - b^n)$ and $a^{n/2} + b^{n/2} \mid (a^n - b^n)$

$a^n = (a^{n/2} \times a^{n/2})$ and $b^n = (b^{n/2} \times b^{n/2})$ therefore $(a^n - b^n) = (a^{n/2} - b^{n/2}) \times (a^{n/2} + b^{n/2}).$

For example:
$(9^{12} - 4^{12}) = (9^6 - 4^6) \times (9^6 + 4^6).$

3.5. Proof of Theorem 2.5: The last digit of an integer has a 2 part cycle for each successive odd power

When an integer is raised from one odd power to the next sequential one,
it is multiplied by its value to the first odd power times itself squared.

For example:
$(5^3 = 5^1 \times 5^2)$, $(5^5 = 5^3 \times 5^2)$ and $(5^7 = 5^5 \times 5^2)$.
This also applies to the integer's last digit. Therefore all ten possible
last digits for an integer, including a mod, either alternate between two

values or remain the same when raised to each successive odd power as follows:

(0, 0) (1, 1) (2, 8) (3, 7) (4, 4) (5, 5) (6, 6) (7, 3) (8, 2) (9, 9).

For example:

$(2^1 = 2, 2^3 = 8), (2^5 = 32, 2^7 = 128), (2^9 = 512, 2^{11} = 2048).$

4. Odd powers > 1

In this section it is demonstrated that, for any two integers (a, b), when $n > 2$ and n is odd, $(a^n - b^n) \notin 10^{1\ldots\infty}$. Separate subsections deal with varying compatible values of $a - b$. Where $a - b$ is incompatible, $(a^n - b^n) \notin 10^{1\ldots\infty}$.

Let us define z as the number of zeroes in $a - b$. Let us define t as the highest value $\in 10^{1\ldots\infty}$ where $t \mid (a - b)$.

(1) When both a and b end in 0:- Let us denote an integer v where $v \in 10^{1\ldots\infty}$. $(a, b) = v \times$ an 'underlying' pair that do not both end in 0.

For example:
$a = 560, b = 400, v = 10$, underlying pair $= 56, 40 \rightarrow (560,\ 400) = 10 \times (56,\ 40)$.

For example:
$(500,\ 400) = 100 \times (5,\ 4)$.

Let us define g as the number of 'trailing' zeroes at the end of a and/or b, whichever has the fewest trailing zeroes . Let $c = \frac{a}{g}, d = \frac{b}{g}$ and $v = 10^g$. $(a^n - b^n) = v^n \times (c^n - d^n)$ therefore, by proving that $(a^n - b^n) \notin 10^{1\ldots\infty}$ for all underlying pairs that do not both end in 0, this will also prove the case for all pairs where a and b both end in 0.

For example:
$a = 5600, b = 4000, g = 2, v = 100, (5600^3 - 4000^3) = 100^3 \times (56^3 - 40^3)$.

(2) When $a - b$ is odd and $n > 1$:- $(a^n - b^n)$ is an odd value $> 1 \rightarrow (a^n - b^n) \notin 10^{1\ldots\infty}$, therefore proof is only required when $a - b$ is an even compatible integer.

4.1. Odd power > 1, (a - b) $\in 2^{1\ldots\infty}$

$a - b$ ends in 2, 4, 6 or 8 therefore a and b must have different last digits. As stated in theorem 1.5, a and b have different last digits to any odd power $\rightarrow (a^n - b^n) \notin 10^{1\ldots\infty}$.

For example:
$(11^3 - 3^3) = (1331 - 27) = 1304$.

4.2. *Power ends in 1, 3, 7 or 9 and EITHER (a and b share a last digit of 1, 3, 7 or 9 and $(a - b) \in \{10^{1\ldots\infty}, (2^{1\ldots\infty} \times 10^{1\ldots\infty}), (5^{1\ldots\infty} \times 10^{1\ldots\infty})\})$ OR (a and b share a last digit of 2, 4, 6 or 8 and $(a - b) \in \{10^{1\ldots\infty}, (2^{1\ldots\infty} \times 10^{1\ldots\infty})\})$*

(1) Let l, r, s represent variable integer values. Any integer x, where $x > 1$, can be defined as $l + r$.

For example:
$x = 34, l = 30, r = 4$. $x^n = n$ pairs of $(l + r)$ multiplied by each other.
For example:
$34^3 = (30 + 4) \times (30 + 4) \times (30 + 4) = 39304$.

$x^n =$ the sum of all left to right permutations of one integer in each pair multiplied together.

For example:
$34^3 = (30 \times 30 \times 30) + (30 \times 30 \times 4) + (30 \times 4 \times 30) + (30 \times 4 \times 4) + (4 \times 30 \times 30) + (4 \times 30 \times 4) + (4 \times 4 \times 30) + (4 \times 4 \times 4)$
$= (1 \times 30^0 \times 4^3) + (3 \times 30^1 \times 4^2) + (3 \times 30^2 \times 4^1) + (1 \times 30^3 \times 4^0)$.

Let us denote p as the first integer in each bracketed set. p is the corresponding value in Pascal's triangle [5]. The total can be expressed as the sum of $n + 1$ values in a loop, $s = 0$ to n where each value = [a variable integer $p \geq 1$] $\times l^s \times r^{n-s}$

There is always exactly one permutation that includes l^0 and exactly one that includes r^0 since there is only one way to order a sequence of identical integers. All other permutations are > 1 since there is more than one way to order non-identical integers.

(2) When a and b have one or more shared right digit(s), $(a^n - b^n)$ can be expressed using two combined values for each integer.
Let us define r as the rightmost digit(s) common to both a and b. Let $u = a - r$ and $v = b - r$.

For example:
$a = 1234, b = 1134, a - b = 100, r = 34, u = 1200, v = 1100$.

$(a^n - b^n)$ equals the sum of $n + 1$ values in a loop, $s = 0$ to n where each value =

$$[\text{a variable integer } p \geq 1] \times (u^s - v^s) \times r^{n-s}$$

u and v both end in one or more trailing zeroes therefore, when $s > 1$, $(u^s - v^s)$ has more trailing zeroes than $u - v$.

EXAMPLE 1. $a = 34, b = 14, r = 4, u = 30, v = 10$

$(34^3 - 14^3) = 36560 =$

$(1 \times (30^0 - 10^0) \times 4^3) = 0$ (always 0, not shown in further examples) +
$(3 \times (30^1 - 10^1) \times 4^2) = 960$ (the same number of trailing zeroes as $a - b$) +
$(3 \times (30^2 - 10^2) \times 4^1) = 9600$ (more trailing zeroes than $a - b$) +
$(1 \times (30^3 - 10^3) \times 4^0) = 26000$ (more trailing zeroes than $a - b$)

When n ends in 1, 3, 7 or 9, a and b share a last digit of 1, 3, 7 or 9 and $(a - b) \in \{ 10^{1...\infty}, (2^{1...\infty} \times 10^{1...\infty}), (5^{1...\infty} \times 10^{1...\infty}) \}$):- the total is an incompatible integer ending in 1, 3, 7 or 9 followed by the same number of trailing zeroes as $a - b$.

For example:
$(113 - 13) = 100$, $(113^3 - 13^3) = 1440700$.

When n ends in 1, 3, 7 or 9, a and b share a last digit of 2, 4, 6 or 8 and $(a - b) \in \{10^{1...\infty}, (2^{1...\infty} \times 10^{1...\infty})\}$:- the total ends in 2, 4, 6 or 8 followed by the same number of trailing zeroes as $a - b$.

For example:
$(34 - 14) = 10$, $(34^3 - 14^3) = 36560$.

4.3. *Power $= 5$, a and b share a last digit of 1, 3, 7 or 9, and $a - b \in$ $\{10^{1\ldots\infty}, (2^{1\ldots\infty} \times 10^{1\ldots\infty}), (5^{1\ldots\infty} \times 10^{1\ldots\infty})\}$*

$(a^5 - b^5)$ is divisible by an incompatible integer that is > 1 and ends in 1.

Proof. $(a^5 - b^5) = (5 \times (u^1 - v^1) \times r^4) +$
$(10 \times (u^2 - v^2) \times r^3) +$
$(10 \times (u^3 - v^3) \times r^2) +$
$(5 \times (u^4 - v^4) \times r^1) +$
$(1 \times (u^5 - v^5) \times r^0)$.
$(u^1 - v^1) = (a - b)$. Each line can be shown after being divided by 5 $\times (a - b)$:

$1 \times 1 \times r^4$ [r ends in 1, 3, 7 or 9 therefore r^4 ends in 1]
$2 \times ((u^2 - v^2)/(a - b)) \times r^3$ [result ends in 0]
$2 \times ((u^3 - v^3)/(a - b)) \times r^2$ [result ends in 0]
$1 \times ((u^4 - v^4)/(a - b)) \times r^1$ [result ends in 0]
$1 \times ((u^5 - v^5)/(a - b))/5 \times r^0$ [result ends in 0].

When $n > 1$, $(\frac{u^n - v^n}{a - b})$ has more trailing zeroes than $a - b$, therefore $(a^5 - b^5) = 5 \times (a - b) \times$ [an incompatible integer ending in 1]. $\qquad \square$

EXAMPLE 2. $a = 37, b = 17, r = 7, u = 30, v = 10$.

$(37^5 - 17^5) = (5 \times (30^1 - 10^1) \times 7^4 = 1 \times (20/20) \times 7^4 \times 5 \times 20 =$
$2401 \times 100 + (10 \times (30^2 - 10^2) \times 7^3 =$
$2 \times (800/20) \times 7^3 \times 5 \times 20 =$
$27440 \times 100) + (10 \times (30^3 - 10^3) \times 7^2 =$
$2 \times (26000/20) \times 7^2 \times 5 \times 20 =$
$127400 \times 100) + (5 \times (30^4 - 10^4) \times 7^1 =$
$1 \times (800000/20) \times 7^1 \times 5 \times 20 =$
$280000 \times 100) + (1 \times (30^5 - 10^5) \times 7^0) =$
$1 \times (24200000/20)/5 \times 7^0 \times 5 \times 20 = 242000 \times 100)$
TOTAL $= 67924100 = (5 \times 20 \times 679241)$.

4.4. *Power* $= 5$, *a and b share a last digit of 2, 4, 6 or 8 and a-b* $\in$ $\{10^{1\ldots\infty}, (2^{1\ldots\infty} \times 10^{1\ldots\infty})\}$

Let $w =$ the highest value $\in \{2^{1\ldots\infty}\}$ where $w \mid a$ and $w \mid b$. Let $c = \frac{a}{w}, d = \frac{b}{w}$.

(1) When c and d share a last digit of 1, 3, 7 or 9:- it was proved in subsection 4.3 that $(a^5 - b^5)$ is incompatible where a and b have a shared ending of 1, 3, 7 or 9, therefore $(c^5 - d^5)$ is incompatible $\to (a^5 - b^5)$ is incompatible since $(a^5 - b^5) = (c^5 - d^5) \times w^5$.

For example:
$a = 92, b = 12, w = 4, c = 23, d = 3$. $(92^5 - 12^5) = (23^5 - 3^5) \times 4^5$.

(2) When $(c - d) \in 5^{1\ldots\infty}$, (for example $a = 48, b = 8, w = 8 \to c = 48/8 = 6, d = 8/8 = 1$. $c - d = 5$):- $(c^5 - d^5)$ is divisible by an incompatible integer > 1 and ending in 1.

Proof. $(c^5 - d^5) = (5 \times (c - d)^1 \times d^4) +$
$(10 \times (c - d)^2 \times d^3) +$
$(10 \times (c - d)^3 \times d^2) +$
$(5 \times (c - d)^4 \times d^1) +$
$(1 \times (c - d)^5 \times d^0)$

Each line can be shown after being divided by $5 \times (c - d)$:
$1 \times 1 \times d^4$ [ends in 6 when d is even or 1 when d is odd]
$2 \times (c - d)^1 \times d^3$ [ends in 0]
$2 \times (c - d)^2 \times d^2$ [ends in 0]
$1 \times (c - d)^3 \times d^1$ [ends in 0 when d is even or 5 when d is odd]
$1 \times ((c - d)^4/5) \times d^0$ [ends in 5, since $25 \mid (c - d)^4$]

Total $=$ an incompatible integer ending in 1. $\qquad\qquad\square$

EXAMPLE 3. $c = 27, d = 2, c - d = 25$.

$(27^5 - 2^5) = (114791 \times 5 \times 25) = (5 \times 25^1 \times 2^4 = 1 \times 1 \times 16 \times 5 \times 25 = 16 \times 5 \times 25) +$
$(10 \times 25^2 \times 2^3 = 2 \times 25 \times 8 \times 5 \times 25 = 400 \times 5 \times 25) +$
$(10 \times 25^3 \times 2^2 = 2 \times 625 \times 4 \times 5 \times 25 = 5000 \times 5 \times 25) +$
$(5 \times 25^4 \times 2^1 = 1 \times 15625 \times 2 \times 5 \times 25 = 31250 \times 5 \times 25) +$
$(1 \times 25^5 \times 2^0 = 1 \times 390625/5 \times 1 \times 5 \times 25 = 78125 \times 5 \times 25)$

4.5. *Power > 5 and ends in 5, and EITHER (a and b share a last digit of 1, 3, 7 or 9 and $a - b \in \{10^{1...\infty}, (2^{1...\infty} \times 10^{1...\infty}), (5^{1...\infty} \times 10^{1...\infty})\}$) OR (a and b share a last digit of 2, 4, 6 or 8 and a-b $\in \{10^{1...\infty}, (2^{1...\infty} \times 10^{1...\infty})\}$)*

For powers 15, 25, 35 etc., $(a^5 - b^5) \mid (a^n - b^n)$. $(a^5 - b^5)$ was proved incompatible in subsections 4.3 and 4.4.

Proof. $(a^n - b^n) = ((a^5 - b^5) \times a^{n-5}) + ((a^{n-5} - b^{n-5}) \times b^5)$. Let $h = n - 5$. Continuously divide h by 2, each time deriving the two values that make up $(a^h - b^h)$, until h ends in a 5.

For example:
$(a^{20} - b^{20}) = ((a^{10} - b^{10}) \times (a^{10} + b^{10}) \rightarrow (a^{10} - b^{10}) = (a^5 - b^5) \times (a^5 + b^5)$.
h starts as 20 and ends as 5. If $h = 5$, it has been proved that $(a^5 - b^5) \mid (a^{n-5} - b^{n-5})$.

If h ends in a 5 and $h > 5$, the process repeats for the new value of h.

For example:
$(a^{70} - b^{70}) = (a^{35} - b^{35}) \times (a^{35} + b^{35}))$. h starts as 70 and ends as 35. $(a^{35} - b^{35})$ is then processed from the start of this subsection. In this way all values of $(a^{n-5} - b^{n-5})$ are proved incompatible by 'descending' down to the incompatible divisor $(a^5 - b^5)$. □

4.6. *Odd power > 1, a and b share a last digit of 2, 4, 6 or 8 and*
$a - b \in \{5^{1\ldots\infty} \times 10^{1\ldots\infty}\}$

Let w = the highest value $\in \{2^{1\ldots\infty}\}$ where $w \mid a$ and $w \mid b$. Let $c = \frac{a}{w}, d = \frac{b}{w}.$.

(1) When c and d share a last digit of 1, 3, 7, or 9:- it was proved in subsections 4.2, 4.3 and 4.5, where n is odd, $n > 1$ and a and b share a last digit of 1, 3, 7 or 9, that $(a^n - b^n)$ is incompatible, therefore $(c^n - d^n)$ is incompatible $\rightarrow (a^n - b^n)$ is incompatible since $(a^n - b^n) = (c^n - d^n) \times w^n$.

For example:
$a = 502, b = 2, w = 2, c = \frac{502}{2} = 251, d = \frac{2}{2} = 1. (502^3 - 2^3) = (251^3 - 1^3) \times 2^3$.

(2) When c and d have different odd last digits:- $(a^n - b^n) \notin 10^{1\ldots\infty}$.

Proof. z is defined as the number of zeroes in $a - b$. The highest value $\in \{2^{1\ldots\infty}\}$ that is divisible into both a and b is 2^z, therefore $w = 2^z$. Where $z >$ the power value of w this falls into part (1) above.
For example:
$a = 2502, b = 2, z = 2, w = 2$, the power value of $w = 1$, therefore $c = 1251, d = 1$.

Let $k = 5^z$. k^n is the only odd value that can be multiplied by w^n to form a value $\in 10^{1\ldots\infty}$. $(a^n - b^n) = (c^n - d^n) \times w^n$.
$c > k$ and $(c - d) > k$ therefore $(c^n - d^n)$ is odd and $(c^n - d^n) > k^n$. $\square$

EXAMPLE 4. $a = 504, b = 4, a - b = 500, z = 2, w = 4, k = 25, c = \frac{504}{4} = 126, d = \frac{4}{4} = 1$.

25^n is the only odd value that can be multiplied by 4^n to form a value $\in 10^{1\ldots\infty}$.

$126 > 25$ and $(126 - 1) > 25$ therefore $(126^n - 1^n)$ is odd and $(126^n - 1^n) > 25^n$.

For example:
$(504^3 - 4^3) = (126^3 - 1^3) \times 4^3$. $(126^3 - 1^3) > 25^3$.

13

4.7. *Odd power > 1, a and b both end in 5 and $a - b \in$*
   $\{10^{1\ldots\infty}, (5^{1\ldots\infty} \times 10^{1\ldots\infty})\}$

$(a^n - b^n) \notin 10^{1\ldots\infty}$.

Proof. z is defined as the number of zeroes in $a - b$.
Let us define y as the highest value $\in 2^{1\ldots\infty}$ where $y \mid (a - b) \rightarrow y = 2^z$,
for example $y = 2$ where $a - b \in \{10, 50, 250\}$ and $y = 4$ where $a - b \in \{100, 500, 2500\}$. a and b have different mods with different last digits when divided by $(y \times 2)$ since $y \mid (a - b)$. Therefore, when n is odd, a^n and b^n have different mods with different last digits when divided by $(y \times 2)$.
$(a - b) > y$ therefore $(a^n - b^n) > y$. $a > 10^z$ and $(a - b) \geq 10^z$ therefore $(a^n - b^n) > 10^z$.
$(y \times 2) \nmid (a^n - b^n) \rightarrow (a^n - b^n) \notin 10^{(z+1)\ldots\infty}$ since $(y \times 2) \mid i$ where $i \in 10^{(z+1)\ldots\infty}$. $\qquad \square$

For example:
$(515 - 15) = 500$. $z = 2$. $y = 4$. $y \times 2 = 8$. $515 > 100$. $(515 - 15) > 100$. $8 \nmid 500$.
$(515^3 - 15^3) = 136587500 = (34146875 \times 4) = (17073437.5 \times 8)$. $8 \mid i$ where $i \in 10^{3\ldots\infty}$.

4.8. *Odd power > 1, a and b both end in 5 and $a - b \in \{2^{1\cdots\infty} \times 10^{1\cdots\infty}\}$*

Let $v =$ the highest value $\in \{5^{1\cdots\infty}\}$ where $v \mid a$ and $v \mid b$. Let $c = \frac{a}{v}$, $d = \frac{b}{v}$.

(1) When c and d share a last digit of 1, 3, 7, or 9:- it was proved in subsections 4.2, 4.3 and 4.5, where n is odd, $n > 1$ and a and b share a last digit of 1, 3, 7 or 9, that $(a^n - b^n)$ is incompatible, therefore $(c^n - d^n)$ is incompatible $\rightarrow (a^n - b^n)$ is incompatible since $(a^n - b^n) = (c^n - d^n) \times v^n$.

For example:
$a = 215$, $b = 15$, $v = 5$, $c = \frac{215}{5} = 43$, $d = \frac{15}{5} = 3$. $(215^3 - 15^3) = (43^3 - 3^3) \times 5^3$.

(2) When c and d have different odd last digits:- $(a^n - b^n) \notin 10^{1\cdots\infty}$.

Proof. z is defined as the number of zeroes in $a - b$. The highest value $\in \{5^{1\cdots\infty}\}$ that is divisible into both a and b is 5^z, therefore $v = 5^z$. Where $z >$ the power value of v this falls into part (1) above.

For example:
$a = 2005, b = 5, z = 2, v = 5$, the power value of $v = 1$, therefore $c = 401, d = 1$.

Let $j = 2^z$. j^n is the only integer ending in 2, 4, 6 or 8 that can be multiplied by v^n to form a value $\in 10^{1\cdots\infty}$. $(a^n - b^n) = (c^n - d^n) \times v^n$. $c > j$ and $(c - d) > j$. c^n and d^n have different odd last digits when n is odd therefore $(c^n - d^n)$ ends in 2, 4, 6 or 8 and $(c^n - d^n) > j^n$. □

For example:
$a = 25, b = 5, z = 1, v = 5, j = 2, c = 25/5 = 5, d = 5/5 = 1$.

2^n is the only integer ending in 2, 4, 6 or 8 that can be multiplied by 5^n to form a value $\in 10^{1\cdots\infty}$. $5 > 2$ and $(5 - 1) > 2$ therefore $(5^n - 1^n) > 2^n$.

For example:
$(25^3 - 5^3) = (5^3 - 1^3) * 5^3$. $(5^3 - 1^3) > 2^3$.

5.   Even powers > 2

In this section it is demonstrated over several subsections that, for any two integers (a, b) when $n > 2$ and n is even, $(a^n - b^n) \notin 10^{1\cdots\infty}$. When (a, b) form an 'underlying' pair with a highest common integer divisor of 1, for example $(17, 11)$, it will be shown that $(a^n - b^n)$ is incompatible when n is even and $n > 2$. All pairs that are not an underlying pair, that is pairs with a highest common integer divisor > 1, are an integer multiple of an underlying pair and have a difference to any power that is a multiple of the underlying pair's difference to that power.

For example:
$(34, 22)$ has underlying pair $(17, 11) \to (34^n - 22^n) = 2^n \times (17^n - 11^n)$.

$(a - b) \mid (a^n - b^n)$ and $(a^{n/2} - b^{n/2}) \mid (a^n - b^n)$, therefore the following subsections relate to cases where either $(a - b)$ or $(a^{n/2} - b^{n/2})$ is compatible and a and b have no shared integer divisor except 1. For even powers, $n \geq 4$. $(a^{n/2} - b^{n/2}) > 1$ when $\frac{n}{2} > 1$, therefore $(a^{n/2} - b^{n/2}) = 1$ will not occur, as demonstrated in subsection 2.5.

5.1.   Even power > 2, $\frac{n}{2}$ is odd and $a - b \in \{2^{1\cdots\infty}\}$

As an underlying pair with no shared divisor except 1, a and b are both odd integers where $a - b$ is an even integer. a and b have different mods with different last digits when divided by $(a - b) \times 2$, therefore when $\frac{n}{2}$ is odd, $a^{n/2}$ and $b^{n/2}$ have different mods with different last digits when divided by $(a - b) \times 2$, as stated in theorem 2.5. Thus $(a^{n/2} - b^{n/2}) > (a - b)$ and $((a - b) \times 2) \nmid (a^{n/2} - b^{n/2})$, therefore $(a^{n/2} - b^{n/2}) \notin 2^{1\cdots\infty} \to (a^{n/2} - b^{n/2})$ forms an incompatible integer > 2 and ending in 2, 4, 6 or 8.

For example:
$a = 11, b = 3$. $a - b = 8$. $(11^3 - 3^3) = 1304 = (163 \times 8) = (81.5 \times (2 \times 8))$.

5.2. Even power > 2, $\frac{n}{2}$ is even and $(a^{n/2} - b^{n/2}) \in 2^{1...\infty}$

$(a^n - b^n)$ is a multiple of $(a^{n/2} + b^{n/2})$, an incompatible value.

Proof. $(a^{n/2} - b^{n/2}) = (a^{n/4} - b^{n/4}) \times (a^{n/4} + b^{n/4})$.
$(a^{n/2} - b^{n/2}) \in 2^{1...\infty}$, therefore $(a^{n/4} - b^{n/4}) \in 2^{1...\infty}$ and
$(a^{n/4} + b^{n/4}) \in 2^{1...\infty}$.
$(a^{n/4} - b^{n/4})$ and $(a^{n/4} + b^{n/4})$ cannot both be divisible by four, since
$a^{n/4}$ and $b^{n/4}$ either share a mod of 1 or 3 when divided by 4 or have
different mods. Therefore $(a^{n/4} - b^{n/4}) = 2$. This is only possible when n
$= 4$ since $(a^{n/4} - b^{n/4}) > 2$ when $n/4 > 1$ as demonstrated in subsection
1.5.

For example:
$(9^4 - 7^4) = (9^2 - 7^2) * (9^2 + 7^2)$. $(9^2 - 7^2) = (9 - 7) \times (9 + 7) = 2 \times 16$.

Let $m = a - 1 \to m = (b + 1)$. $m = (\frac{a+b}{2}) \to m \in 2^{1...\infty}$.

For example: $a = 5, b = 3, m = 4$.
$a^2 = (m^2 + (2 \times m)) + 1$ and $b^2 = (m^2 - (2 \times b)) - 1$ therefore $(a^2 + b^2)$
$= (2 \times m^2) + (2 \times (m - b)) = (2 \times m^2) + (2 \times 1) = 2 \times (m^2 + 1)$

$m \in 2^{1...\infty} \to m^2$ ends in 2, 4, 6 or 8.

When $(m^2 + 1)$ ends in 3, 7 or 9 it is an incompatible integer.

When $(m^2 + 1) = 5$:- $a = 3, b = 1$ and $m = 2$. $(3^4 - 1^4) = 80$. 80 cannot
be multiplied by any integer to the power of 4 to form a value $\in 10^{1...\infty}$
since 80 must be multiplied by $(5^3 \times i)$ where $i \in 10^{0...\infty}$. $(3^4 + 1^4) = 82$.
82 is incompatible therefore $(3^8 - 1^8)$ is also incompatible.

When $(m^2 + 1) > 5$ and $(m^2 + 1)$ ends in 5:- $(m^2 > 4)$, $(8 \mid m^2)$ and
$(6 \nmid m^2) \to (m^2 + 1) \notin 5^{2...\infty}$ since 5 to any odd power has a mod of 5
when divided by 6 or 8 and 5 to any even power has a mod of 1 when
divided by 6 or 8. That is $5 \times$ the starting mod of $5 = 25 = $ a new mod of
1 when divided by 6 or 8 and $5 \times$ mod $1 = $ mod 5. Therefore deducting
1 from any value $\in 5^{2...\infty}$ results in integer i where $(6 \mid i)$ or $(8 \nmid i)$. $\square$

EXAMPLE 5. $a = 5, b = 3, m = 4$.
$5^2 = (4^2 + (2 \times 4)) + 1$ and $3^2 = (4^2 - (2 \times 3)) - 1$ therefore $(5^2 + 3^2) = $
$(2 \times 4^2) + (2 \times (4-3)) = (2 \times 4^2) + 2 = 2 \times (16 + 1)$
Note: $m^2 + 1 = 17$

5.3. Even power > 2, $(a^{n/2} - b^{n/2}) \in \{5^{1...\infty}\}$

All powers of 5 end in 5, therefore $(a^{n/2} - b^{n/2})$ ends in a 5. This happens when $a^{n/2}$ and $b^{n/2}$ have one of four pairs of last digits, either way round: $(9, 4)$, $(8, 3)$, $(7, 2)$, $(6, 1)$. As an underlying pair with no shared divisor except 1, a and b will not have the last digits $(5, 0)$, either way round. Therefore $(a^{n/2} + b^{n/2})$ is an incompatible integer > 1 and ending in 3, 1, 9 or 7.

For example:
$(13^4 - 12^4) = (13^2 - 12^2) \times (13^2 + 12^2) = (169 - 144) \times (169 + 144) = 25 \times 313$.

5.4. Even power > 2, $\frac{n}{2}$ is odd and $(a^{n/2} - b^{n/2}) \in \{10^{1...\infty}, (2^{1...\infty} \times 10^{1...\infty}), (5^{1...\infty} \times 10^{1...\infty})\}$

As an underlying pair with no shared divisor except 1, a and b are both odd integers where $(a^{n/2} - b^{n/2})$ is an even integer. Therefore $a^{n/2}$ and $b^{n/2}$ share a last digit of 1, 3, 7 or 9. $(a^{n/2} - b^{n/2})$ is already proved in subsections 4.2, 4.3 and 4.5 to be incompatible for all odd powers > 1 when a and b share a last digit of 1, 3, 7 or 9, for example $(23^3 - 3^3)$, therefore $(a^n - b^n)$ is also incompatible.

For example: $(23^6 - 3^6) = (23^3 - 3^3) \times (23^3 + 3^3)$.

5.5. Even power > 2, $\frac{n}{2}$ is even and $((a^{n/2} - b^{n/2}) = 10$ or $(a^{n/2} - b^{n/2}) \in \{5^{1\ldots\infty} \times 10\})$

When $(a^{n/2} - b^{n/2}) \in \{5^{1\ldots\infty} \times 10\}$, $(a^{n/2} - b^{n/2})$ ends in 50. Integers with a mod of 1 when divided by 4 (1, 5, 9 etc.) have a mod of 1 when raised to any power and divided by 4. For integers with a mod of 3 (3, 7, 11 etc.) this mod becomes 1 on an even power, that is $3 \times 3 = 9 =$ mod 1, then alternates back to 3 on the next odd power. All odd integers therefore have a mod of 1 when raised to any even power and divided by 4, thus $4 \mid (a^{n/2} - b^{n/2})$. $(a^{n/2} - b^{n/2})$ therefore cannot $= 10$ or end in 50 since these values would only occur if $4 \nmid (a^{n/2} - b^{n/2})$.

5.6. Even power > 2, $(a^{n/2} - b^{n/2}) \in \{10^{2\ldots\infty}, (2^{1\ldots\infty} \times 10^{1\ldots\infty}), (5^{1\ldots\infty} \times 10^{2\ldots\infty})\}$

$a^{n/2}$ and $b^{n/2}$ share a last digit of 1, 3, 7 or 9. $4 \mid (a^{n/2} - b^{n/2})$ therefore $a^{n/2}$ and $b^{n/2}$ share a mod of 1 or 3 when divided by $4 \to (a^{n/2} + b^{n/2})$ has a total mod of 2 when divided by 4. $(a^{n/2} + b^{n/2})$ is therefore an incompatible integer > 2, since $4 \nmid (a^{n/2} + b^{n/2})$ and $(a^{n/2} + b^{n/2})$ ends in 2, 4, 6 or 8.

For example:
$(633^4 - 617^4) = (633^2 - 617^2) \times (633^2 + 617^2) = 20000 \times 781378$ ($4 \nmid$ 781378)

6. Concluding remarks

It has been proved, using the principle of ratios, that the sum of any two integers raised to any power $n > 2$ is a total that has an irrational nth root. This proves Fermat's Last Theorem.

7. Appendix 1: Coverage of even compatible values for odd powers > 1

$a - b$	Last digits of A and B	Odd powers	Section
$2^{1\ldots\infty}$	All	All	4.1
$10^{1\ldots\infty}$	Shared LD of 1, 3, 7 or 9	Power ends in 1, 3, 7 or 9	4.2
$(2^{1\ldots\infty} \times 10^{1\ldots\infty}), (5^{1\ldots\infty} \times 10^{1\ldots\infty})$	Shared LD of 1, 3, 7 or 9	Power ends in 1, 3, 7 or 9	4.2
$10^{1\ldots\infty}$	Shared LD of 1, 3, 7 or 9	Power = 5	4.3
$(2^{1\ldots\infty} \times 10^{1\ldots\infty}), (5^{1\ldots\infty} \times 10^{1\ldots\infty})$	Shared LD of 1, 3, 7 or 9	Power = 5	4.3
$10^{1\ldots\infty}$	Shared LD of 1, 3, 7 or 9	Power > 5 and power ends in 5	4.5
$(2^{1\ldots\infty} \times 10^{1\ldots\infty}), (5^{1\ldots\infty} \times 10^{1\ldots\infty})$	Shared LD of 1, 3, 7 or 9	Power > 5 and power ends in 5	4.5
$10^{1\ldots\infty}, (2^{1\ldots\infty} \times 10^{1\ldots\infty})$	Shared LD of 2, 4, 6 or 8	Power ends in 1, 3, 7 or 9	4.2
$10^{1\ldots\infty}, (2^{1\ldots\infty} \times 10^{1\ldots\infty})$	Shared LD of 2, 4, 6 or 8	Power = 5	4.4
$10^{1\ldots\infty}, (2^{1\ldots\infty} \times 10^{1\ldots\infty})$	Shared LD of 2, 4, 6 or 8	Power > 5 and power ends in 5	4.5
$(5^{1\ldots\infty} \times 10^{1\ldots\infty})$	Shared LD of 2, 4, 6 or 8	All	4.6
$10^{1\ldots\infty}, (5^{1\ldots\infty} \times 10^{1\ldots\infty})$	Shared LD of 5	All	4.7
$(2^{1\ldots\infty} \times 10^{1\ldots\infty})$	Shared LD of 5	All	4.8

8. Appendix 2: Coverage of all compatible values for even powers > 2

Value used	Compatible value	$n/2$	Section
a-b	$2^{1\ldots\infty}$	Odd	5.1
$a^{n/2} - b^{n/2}$	$2^{1\ldots\infty}$	Even	5.2
$a^{n/2} - b^{n/2}$	$5^{1\ldots\infty}$	All	5.3
$a^{n/2} - b^{n/2}$	$10^{1\ldots\infty}, (2^{1\ldots\infty} \times 10^{1\ldots\infty}), (5^{1\ldots\infty} \times 10^{1\ldots\infty})$	Odd	5.4
$a^{n/2} - b^{n/2}$	10 only, $(5^{1\ldots\infty} * 10)$	Even	5.5
$a^{n/2} - b^{n/2}$	$10^{2\ldots\infty}, (2^{1\ldots\infty} \times 10^{1\ldots\infty}), (5^{1\ldots\infty} * 10^{2\ldots\infty})$	Even	5.6

Acknowledgements. The author wishes to thank Michael Kielstra for his extremely thorough and helpful review of this paper.

References

1. SIMON SINGH, 'Fermat's Last Theorem', (Fourth Estate, London, 1997) ISBN 1-85702-521-0.
2. A. WILES, 'Modular Elliptic Curves and Fermat's Last Theorem. Annals of Mathematics', (1995) 443-551.
3. R. TAYOR AND A. WILES, 'Ring theoretic properties of certain Hecke algebras. Annals of Mathematics', (1995) 553-572.
4. M. VOS SAVANT, 'The World's Most Famous Math Problem: The Proof of Fermat's Last Theorem and Other Mathematical Mysteries. St Martin's Press', (1993) ISBN-10: 0312106572. ISBN-13: 978-0312106577.
5. A.W.F. EDWARDS, 'Pascal's Arithmetical Triangle: The Story of a Mathematical Idea. Johns Hopkins Paperback', (2002) ISBN-10: 0801869463. ISBN-13: 978-0801869464.

Paul Mansbridge, Croydon, UK.

emailinbox@blueyonder.co.uk

www.ingramcontent.com/pod-product-compliance
Lightning Source LLC
Chambersburg PA
CBHW070735180526
45167CB00004B/1771